Let's find out about
the Moon

by MARTHA and CHARLES SHAPP

Pictures by YUKIO TASHIRO

SCHOLASTIC BOOK SERVICES

NEW YORK • LONDON • RICHMOND HILL, ONTARIO

1st printing .... November 1966
Printed in the U.S.A.

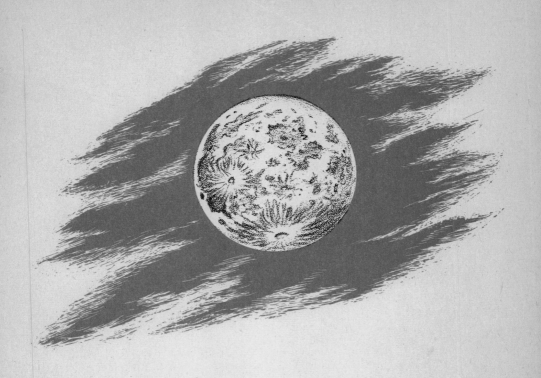

The moon is our nearest neighbor in space.

It seems bigger and brighter than the stars, but it is really much smaller.

The moon is a very small planetary body. It seems bigger than the stars because it is so much closer to us than any star.

The moon travels around the earth just as the earth travels around the sun.

People have always been interested in the moon.

But long, long ago, people had no way to find out about the moon. They just looked at it and wondered about it.

Then the telescope was invented.

Scientists looked at the moon through telescopes. They were able to study the moon, and they learned many things.

In the last few years, spacecraft have been sent to the moon. Some of them have sent back pictures of the moon by television.

Scientists now know a great deal about the moon. But they want to know more.

8

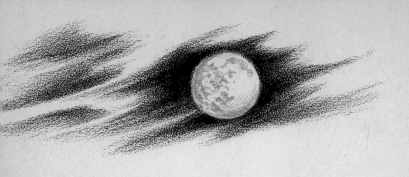

The best way to find out more about the
moon is for someone to go and see.

So the scientists are getting ready to blast
off a rocket that will send men to the moon.

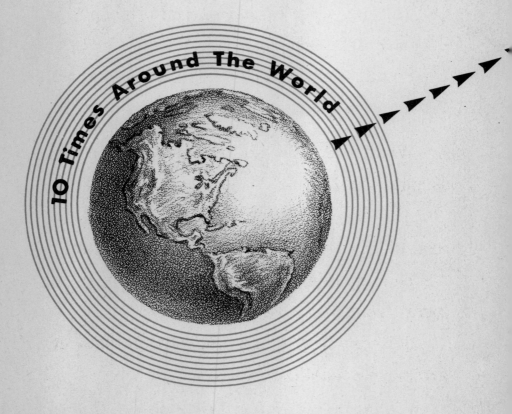

10 Times Around The World

What a difficult and dangerous trip that is going to be!

Although the moon is closer to us than any of the stars, it isn't really very close.

The moon is about 240 thousand miles away from our earth.

This is almost ten times the distance around our earth.

65 MILES AN HOUR

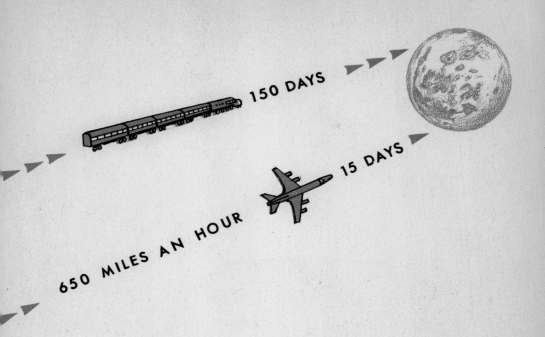

A train speeding day and night at 65 miles an hour would take about 150 days for the trip. This is about five months.

A jet plane speeding day and night at 650 miles an hour would take about 15 days. (But of course a jet can't fly in outer space anyhow.)

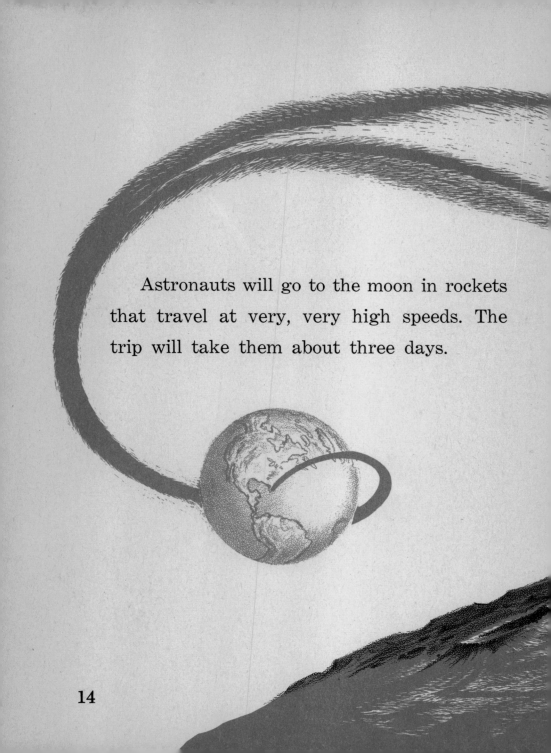

Astronauts will go to the moon in rockets that travel at very, very high speeds. The trip will take them about three days.

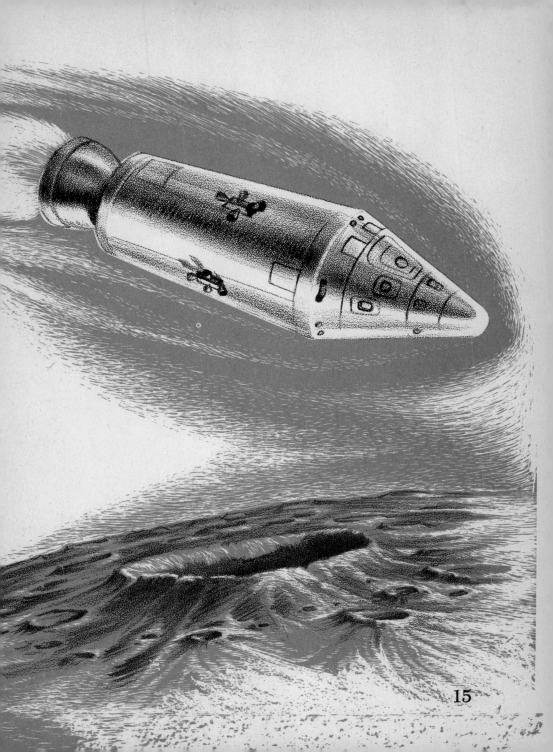

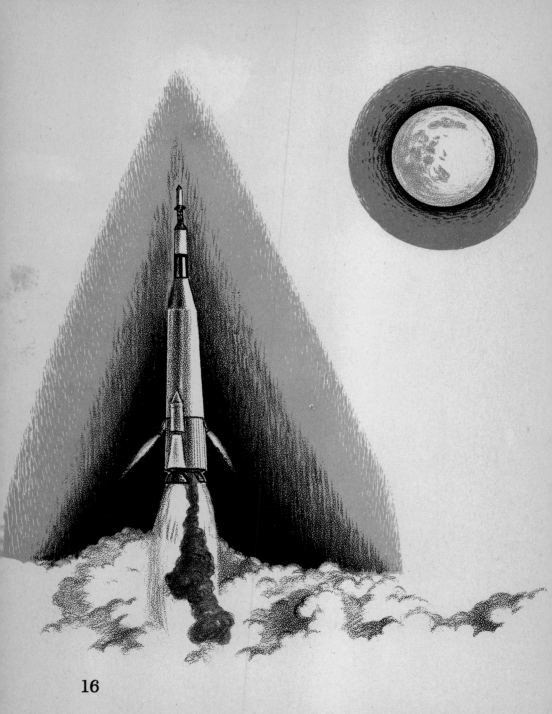

The moon rockets must be shot from the earth.

Imagine shooting at a target 240 thousand miles away! And that target is quite small.

If the earth were the size of a basket-ball, the moon would be the size of a tennis ball.

2,300 MILES AN HOUR

The moon is a small target to hit. And it is a moving target, too.

The moon circles around the earth at a speed of about 2,300 miles an hour.

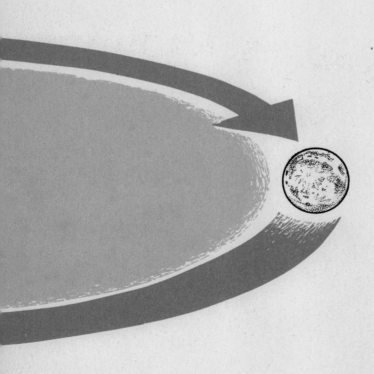

When the astronauts land on the moon, what will they find?

They will find a silent world.

They will see high, jagged mountains and deep holes called craters.

The astronauts will need special machines to travel over this rough surface.

The astronauts will find no water on the moon's surface.

People cannot live without water.

The astronauts will have to bring water from the earth or make it in their spacecraft.

The astronauts will find no air on the moon.

People need air to breathe.

The astronauts will have to bring a supply of air. They will have to carry air tanks with them all the time.

The astronauts will find the moon a very silent place.

They will not hear a sound.

Sound is carried by air. Since there is no air on the moon, there is no sound.

How will the astronauts talk to each other?

They will have to talk through radios. Radio waves can travel without air.

There are days and nights on the moon.
But each day and each night is two weeks
long.

During the two weeks of daytime, the
heat from the sun is so great that human
beings could not live.

To work on the moon in the daytime,
astronauts will need cooling systems built
into their clothes.

During the two weeks of night on the moon, it is so cold that a man would quickly freeze to death.

To go out into the night, an astronaut would need a heating system built into his clothes.

The force of gravity on the moon is weaker than it is on earth.

This means that things are lighter on the moon than they are on earth.

If you weigh 60 pounds on earth, you would weigh only 10 pounds on the moon.

If you can jump four feet on earth, you would be able to jump 24 feet on the moon.

How does the moon look to us here on earth?

It looks like a big shiny ball.

But the moon does not shine with its own light. The moon gets its light from the sun, just as the earth does.

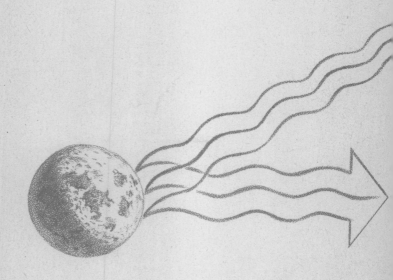

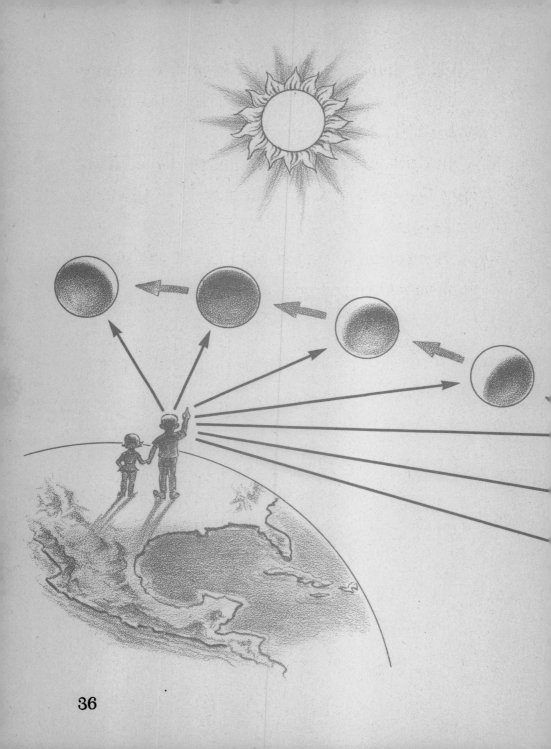

The shape of the moon seems to change a little from night to night. But the moon doesn't really change shape.

As the moon circles the earth, we can see only the part of the moon that is lighted up by the sun.

Sometimes we see a big part of the moon. Sometimes we see only a little part.

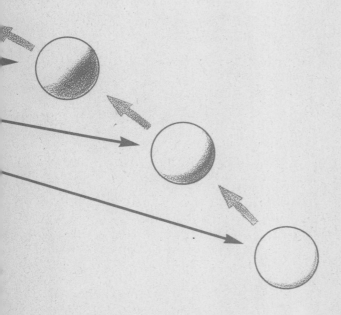

Sometimes the moon looks big and round.
Sometimes it looks like half a ball.
Sometimes we see only a slice of it.
And sometimes we do not see it at all.

When the moon comes between the earth and the sun, we cannot see the moon. Why? Because the side of the moon facing the earth is not lighted up by the sun.

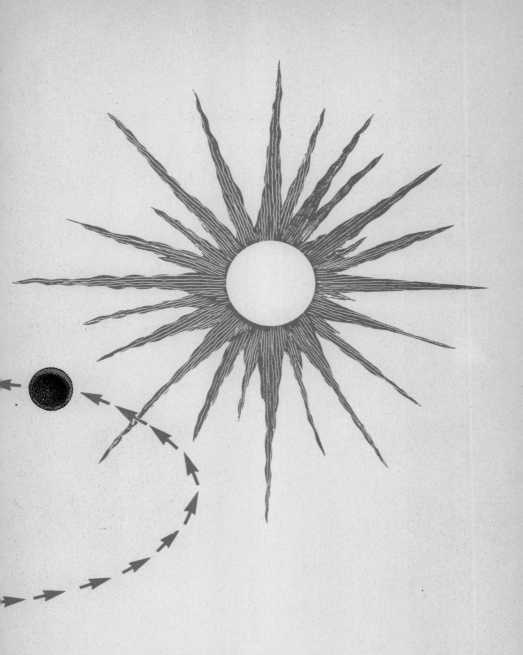

It takes the moon about a month to go completely around the earth.

As it travels, we see a little more of the moon.

Each night for two weeks we see more and more of the lighted side of the moon.

The moon has made half its trip around
the earth in two weeks.

Now the earth is between the moon and
the sun.

We see one whole side of the moon lighted
up by the sun. We call this the full moon.

For the next two weeks, we see less and less of the lighted side of the moon. Then we cannot see it at all.

The same thing happens the next month and the next month and the next month.

Maybe some day, after men reach the
moon, they will blast off from a moon station
and reach another planet.